CONTENTS

TWO TYPES OF ENERGY

▲ Twigs burst into flames when they touch the hot rock on the island of Lanzarote. Here, the underground hot rock comes to the surface.

Geothermal energy

Geothermal energy is heat deep underground. The centre of the Earth is very hot. It is so hot that it melts the rock around it. A lump of this super-hot rock, about the size of a mountain, has enough heat energy in it to power the whole world for a year. If we could turn more of this hot energy into electricity, it would be very useful.

Geothermal power around the world

About 25 countries use geothermal power. The largest users are the USA and the Philippines. Other users of geothermal power include New Zealand, Russia, Mexico, Italy, Japan, Indonesia and Turkey.

But geothermal power only provides 0.15 per cent of the world's electricity. Most electricity is produced by burning fossil fuels (coal, oil and gas) in power stations.

▲ A geothermal power station in Iceland. Hot rocks underground heat water and make steam. The steam turns turbines that power generators, to make electricity. Then the steam cools down to make hot water that is put into a lake. The water is hot enough for people to bathe in it.

FACTFILE

WHAT'S A WATT?
Scientists measure electrical energy in joules. Using one joule per second is called one watt, for example a 40 watt light bulb uses 40 joules for every second it is switched on.

A kilowatt = one thousand watts.
A megawatt = one million watts.
A gigawatt = a thousand million watts.

Bio-energy

Bio-energy is power produced from plants and animals. Enough plants grow every year to meet the world's energy needs eight times over. Also, plants produce seeds so they renew themselves, unless we use them up too fast.

Green plants make their own food. They use:

- sunlight (solar energy)
- water (from the soil)
- carbon dioxide (from the air).

We can burn plants for light, heat and cooking. We can make clothes and tools from plants. We can even build houses from plants. ▼

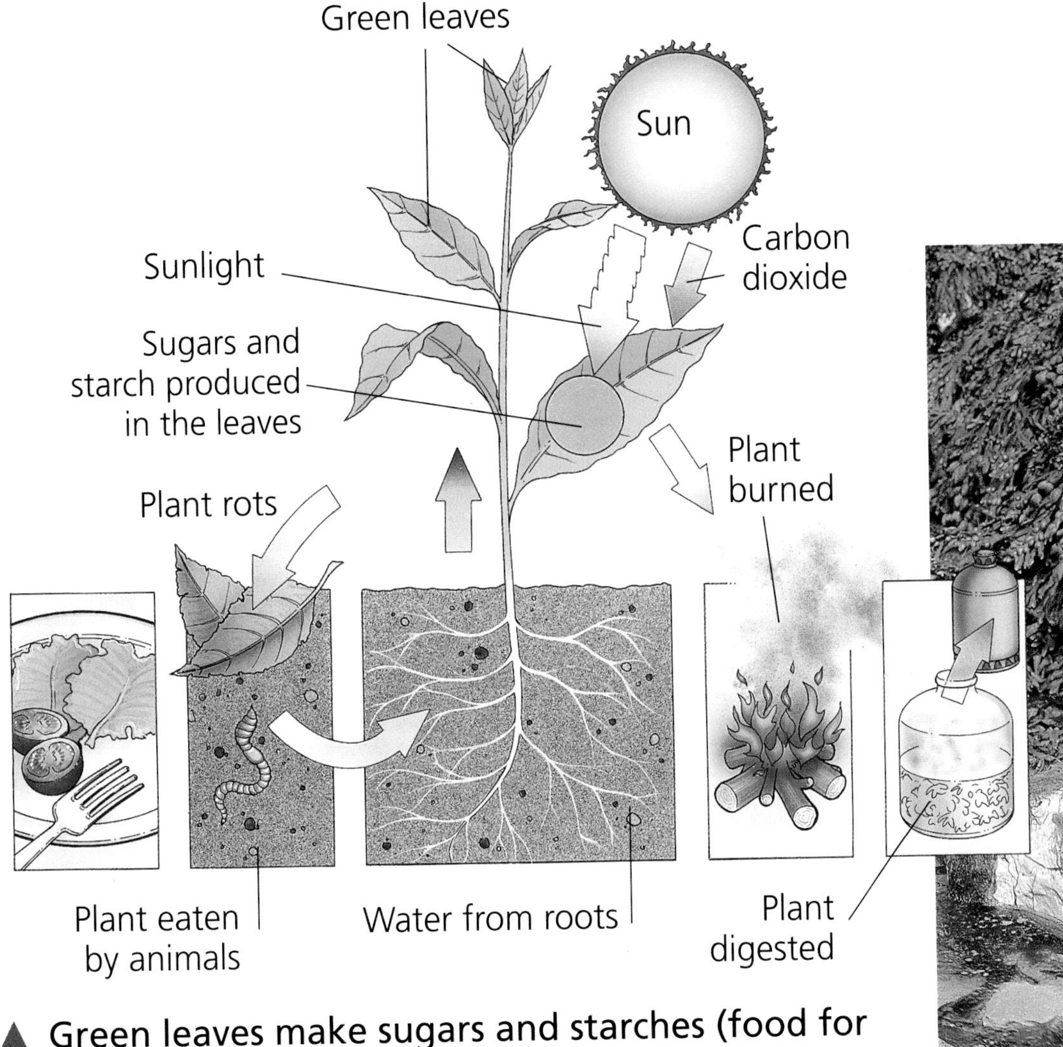

▲ Green leaves make sugars and starches (food for the plant) by using sunlight, carbon dioxide and water.

How a plant makes its food

A plant uses solar energy, water and carbon dioxide to make food. This is called photosynthesis. The plant can then use the food. We can eat plants and so use their energy. We can also burn plants to give us energy as heat. By burning plants in power stations we can change their energy into electricity. It would be useful if we could make electricity directly from plants.

▲ Seaweed is a very common plant. But so far, it has not been used very much in power stations, to make electricity.

GEOTHERMAL AND BIO-ENERGY

The Earth is made up of:

- a solid metal inner core
- a liquid metal outer core
- a thick mantle which moves like warm toffee
- a thin, solid, cool crust of rock.

Everything is very hot (except the crust with the continents and oceans on top of it). ▼

What does geothermal mean?

The word geothermal comes from two Greek words:

geo = earth   *therme* = heat

So it is used to describe heat from the Earth.

Where does the heat come from?

Most of the energy that heats the Earth's core comes from nuclear reactions taking place inside the Earth. Luckily there is a cool, outer crust for us to live on.

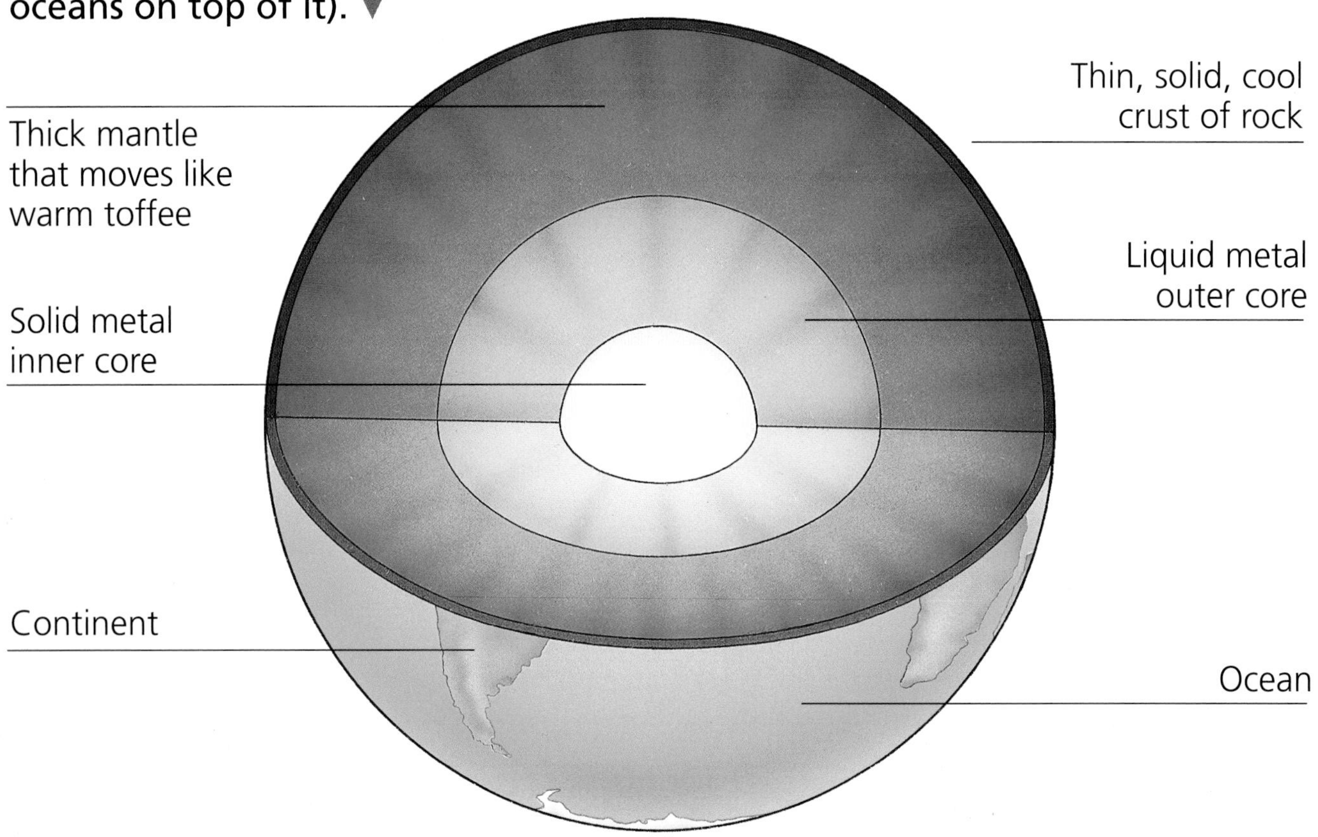

Geothermal heating

Heat comes up from the core. But it gets cooler and cooler as it nears the surface crust. When it reaches the crust the heat is let off or radiated into space. We do not notice the geothermal heating, because the heat from the Sun is greater.

FACTFILE

The Earth's crust that we live on is like huge pieces of crazy paving. Each piece of crazy paving is called a plate. The plates move slowly over the mantle. They carry the continents on top of them.

Tenerife is a Spanish island. It is made of ash and lava from volcanoes. Volcanoes erupt and bring up red hot material from deep underground.

◀ This lichen is growing on a tree trunk. Lichens grow all over the world. Local people often burn them as fuel for heating and cooking.

Where do bio-fuels come from?

Bio-fuels come from:

- trees
- shrubs
- grasses
- peat from bogs
- seaweed
- mosses and lichens
- animal dung (often contains bits of plants).

These bio-fuels are all around us. Huge forests cover parts of North America, Northern Europe and Asia. Other forests cover large parts of South America, Africa and the Far East. Apart from forests, there are large areas of grassland and shrubs in the world.

This is a photograph of forest and grassland in a cold part of Sweden. Even in cold places, some plants grow well. They could be used to make fuel. ▼

◀ These are pine logs. The wood will be used for building. The waste wood chips and bark can be burnt as fuel to make electricity.

The effects of geothermal energy

Geothermal energy drives the very slow movement of the continents (see the Factfile on page 9). The movement of the continents is called continental drift. As the continents drift they knock into each other, which can cause mountains to be pushed up. If the continents pull apart instead, huge valleys can be formed. All this happens over many millions of years.

Volcanoes show that geothermal energy can also move fast. Volcanoes erupt. They throw up red-hot lava, gas and ash from the super-hot rock inside the Earth.

FACTFILE

There are about 850 active volcanoes in the world today. Most of them are in the Pacific Ocean.

◀ This photograph shows rivers of red-hot lava on the islands of Hawaii in the Pacific Ocean. The islands were made from volcanoes that grew up from the ocean floor.

Snow-capped volcanoes

Volcanoes can form high mountains, so they can have snow on the top. If the volcano erupts, the snow mixes with the red-hot lava and ash. It makes a river of mud. This can flow down the mountain and wash whole villages away.

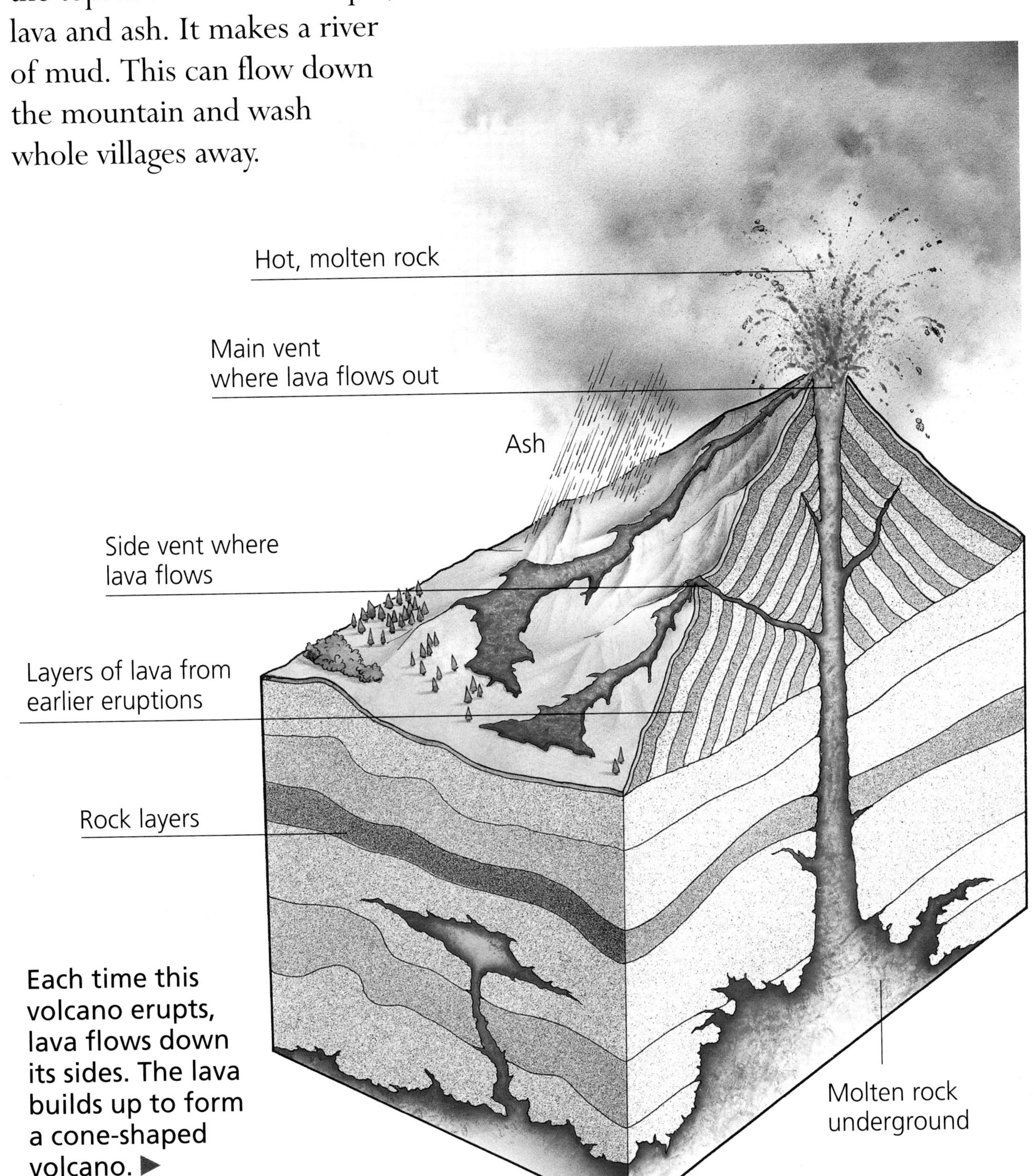

Each time this volcano erupts, lava flows down its sides. The lava builds up to form a cone-shaped volcano. ▶

GEOTHERMAL AND BIO-ENERGY IN HISTORY

Wood-burning bio-energy

For thousands of years people have burned wood. This has kept them warm and cooked their food. In many developing countries, wood fires are still the only way of cooking and heating.

Rubbish and gas bio-energy

In some countries rubbish is burned to make electricity. Also, gas from rotting plants has been used as a fuel.

A family in north-west India sits round a wood fire. Wood is the most important fuel for millions of people. ▼

Geothermal power

In some places underground water is heated by geothermal energy. The water comes to the surface as a hot spring (see page 30). The Romans built bath houses wherever they found hot springs. They enjoyed having hot baths and thought the minerals in the water were good for their health. They built hot spring bath houses all across the Roman empire.

This is an 18th century picture. It shows hot spring baths. Hot springs often have minerals in them, and people thought these were good for them.They sat in the water for hours, or breathed in the steam.

What is peat?

Mosses grow well in wetlands (bogs). As they die, they sink. Then they are squashed by new mosses growing on the top. Slowly the dead mosses build up in layers. They form what is called peat. There are peat bogs all over the world. About one-third of all of Finland is peat bog. Canada, Ireland and Russia all have a lot of peat.

Peat bogs can be many metres thick. The peat can be cut into bricks. It can be burnt on fires or in power stations to make electricity. ▼

◀ This photograph shows piles of hand-cut blocks of peat. In the past, peat was always dug by hand.

Peat as a fuel

Nowadays peat is used in power stations to make electricity for the following places:

- most of Finland's inland cities
- about one fifth of homes and factories in the Irish Republic
- some parts of Russia
- some parts of the USA.

FACTFILE

Things do not rot much in peat bogs. Animals and humans who fell into bogs thousands of years ago are still well preserved. Hundreds of human bodies dating from over 2,000 years ago have been found in bogs all over Europe.

◀ Peat is now dug out by machines. Apart from fuel its uses have included bricks for building, and mopping up oil spills.

GEOTHERMAL AND BIO-TECHNOLOGY

Hot wells

Hot rocks lying underground can be reached by drilling. Then underground water flowing over the rock will be heated. The hot water can be brought up to the surface and used to make electricity.

The geothermal power station makes,or generates, electricity from the hot water that comes up the second hole from underground. ▼

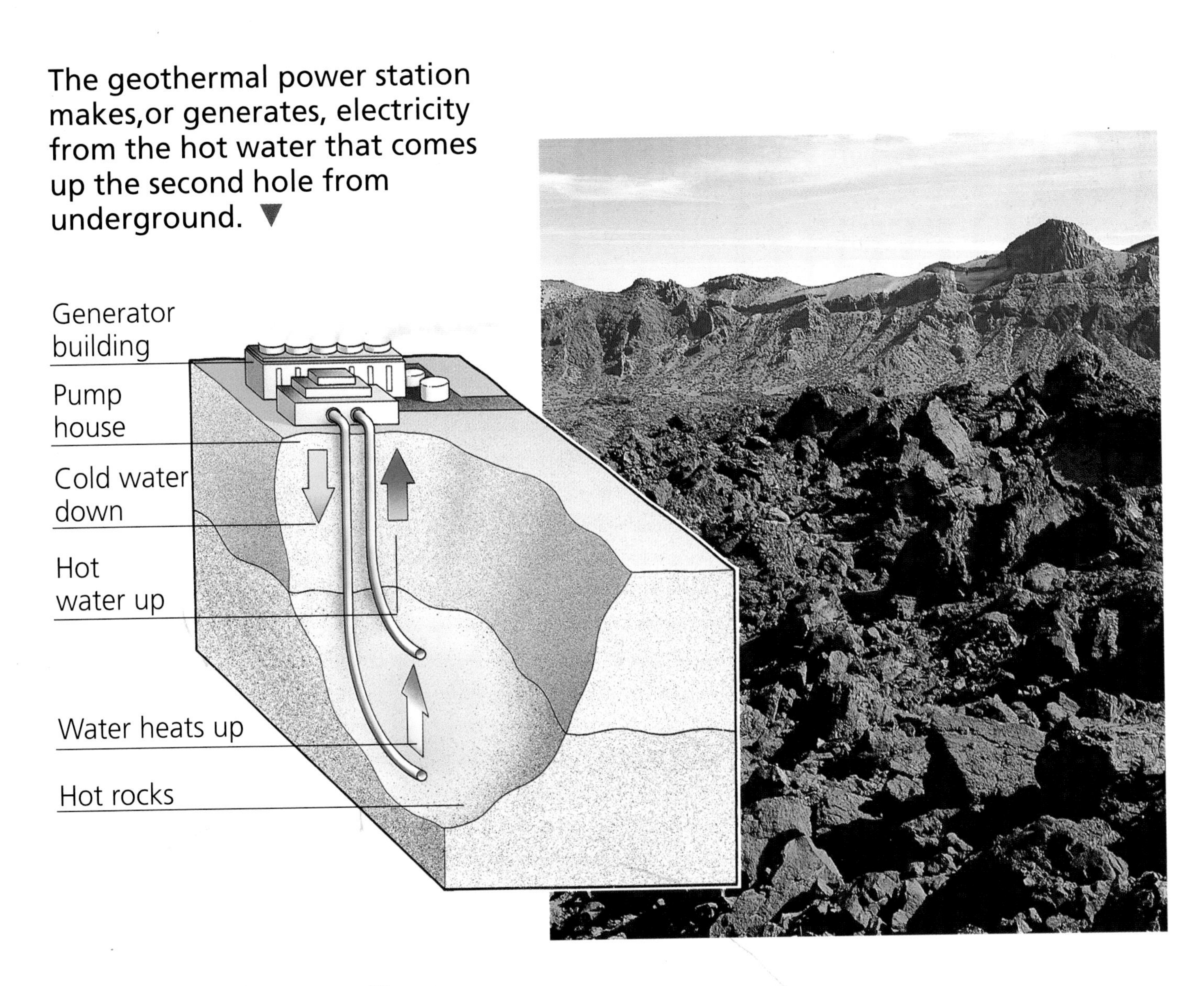

▲ These lava fields on the island of Tenerife show that there was once a volcano here. The volcano may be so old that the hot rock is far below the surface now.

Making a hot well

The hot rocks underground may be dry. So this is what engineers do.

- They drill two holes.
- Cold water is pumped down the first hole.
- The hot underground rocks heat the water up.
- Then the hot water comes back up the second hole (see the diagram above).

◀ A muddy pool heated by underground rocks, in Iceland.

▲ This photo shows a geothermal power station in California, in the USA.

Los Alamos National Laboratory

The Los Alamos National Laboratory is in New Mexico, USA. In 1986, two wells were drilled into the rock there.

Cold water was pumped down a hole into a big, natural space in the rock (see the diagram on page 21). The hot rocks under the ground heated the water. The hot water, or steam, was forced up the second hole. It was about 190 degrees Celsuis when it came to the surface.

In the power station

The power station was on the surface, above the pipes. In the power station, the heat was taken out of the water. It was used to make electricity. Then the cooled water was pumped back down the first hole, and heated up again.

FACTFILE

The Los Alamos National Laboratory is in the USA. It was set up in 1943. It carries out research into weapons and many other things. This includes finding new ways of making electricity.

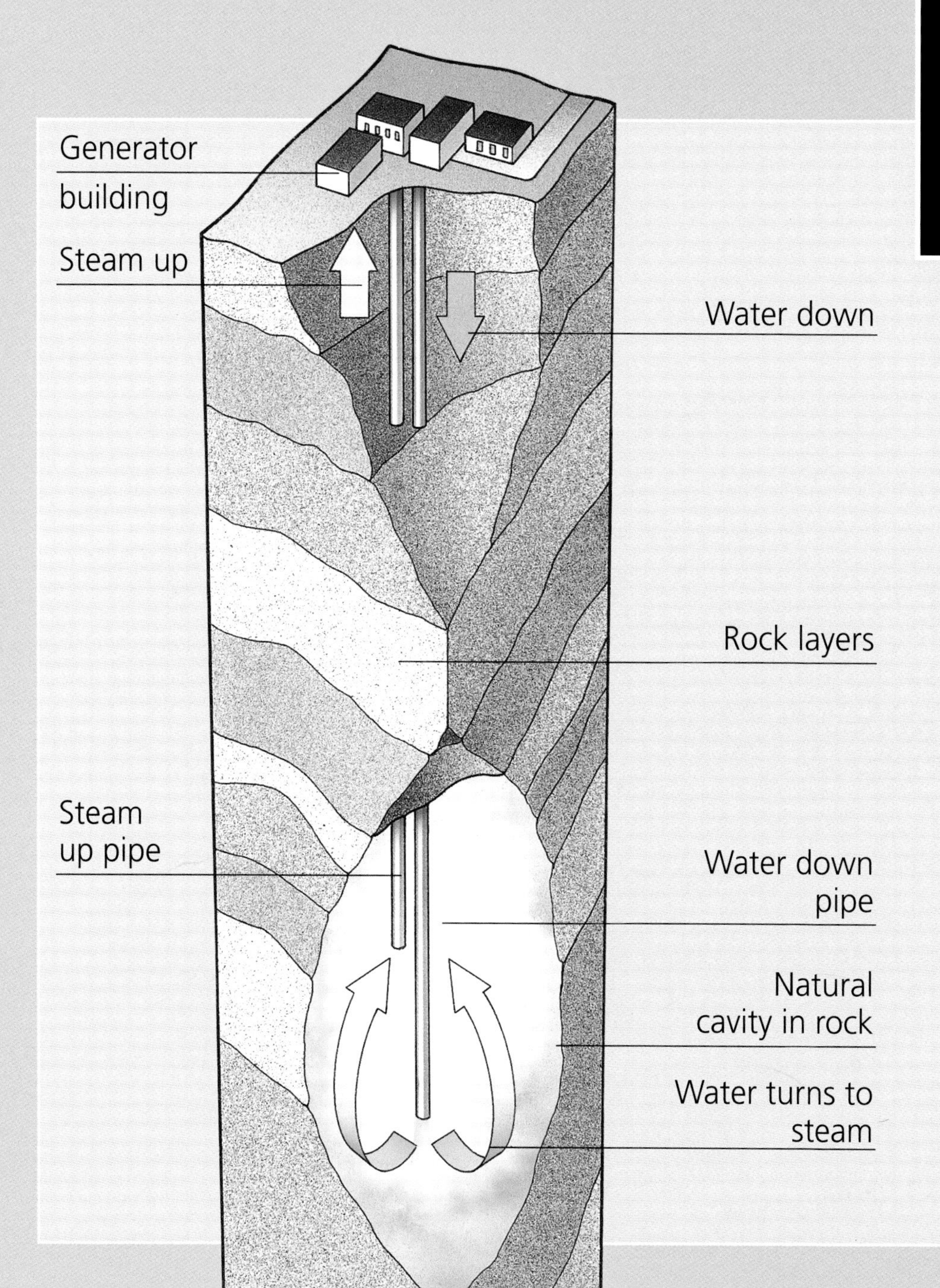

◀ This experiment at Los Alamos showed that water could be used to bring heat up from underground, and then turn it into electricity.

FACTFILE

Waste can be sorted. Some can be recycled. Dangerous materials can be removed. The rest, which can be burnt, is called Refuse Derived Fuel. This is mostly plant material, paper and plastic. RDF can be burnt to make electricity. There is a power station in London which makes electricity in this way.

Burning

In some countries there is plenty of wood, so wood is burnt to warm houses and for cooking. Often in saw mills and paper mills, waste bits of wood are burnt to heat the building, dry new timber and to make electricity (see the diagram below). We can burn other rubbish in special containers that do not allow the air around to be polluted.

Burying

There is quite a lot of plant material in everyday household rubbish (often about one third). Household rubbish has always been buried on waste ground. But now there is so much we are running out of places to bury it.

MAKING ELECTRICITY IN A SAWMILL
Waste bits of wood are burnt in a furnace. This heats water to make steam. The steam drives a turbine. The turbine powers a generator which makes electricity. ▶

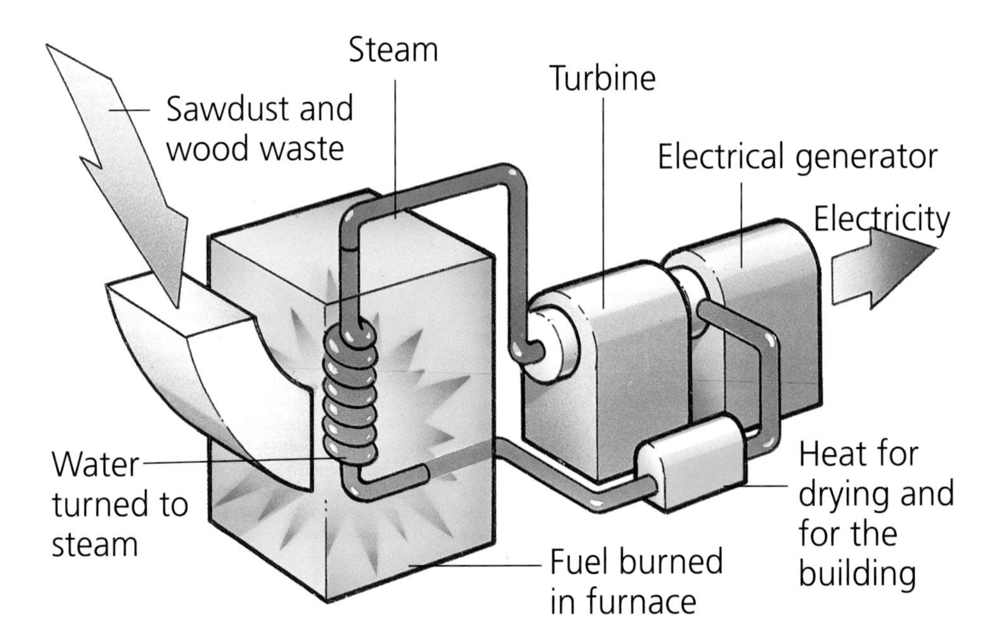

▲ Straw from crops such as wheat can be burnt to make heat.

◀ This power station in California, in the USA, burns wood and farming rubbish to make electricity.

Animal dung

Bacteria are tiny bugs. They are so tiny that you cannot see them unless you look through a microscope. Some bacteria grow well in covered tanks. These bacteria are used to rot down animal dung. This makes a gas.

In India and China, people often collect animal dung to use as fuel. They put it into tanks. Bacteria break down the dung and gas is given off. This gas can be used for cooking, heating and even driving generators to make electricity.

Hundreds of thousands of animals are kept for farming in Africa and Asia. Their dung can be rotted to make gas, or dried to burn as a fuel. ▼

Gas from waste or dung

The gas produced from rotting waste or dung is called methane. Some methane gas is collected and used to make electricity in Britain. But methane gas is only a small part of all Britain's electricity supply.

- About 2 per cent comes from methane collected at human sewage works.
- About 5 per cent comes from methane collected from rotting rubbish. Burning rubbish also gives about 7 per cent.

FACTFILE

The daily waste of one chicken and one human being produces:

- chicken = 14 litres of methane gas.
- human being = 28 litres of methane gas

This photograph shows methane being made at a sewage works. ▶

This diagram shows methane being made by bacteria from slurry (liquid animal dung). ▼

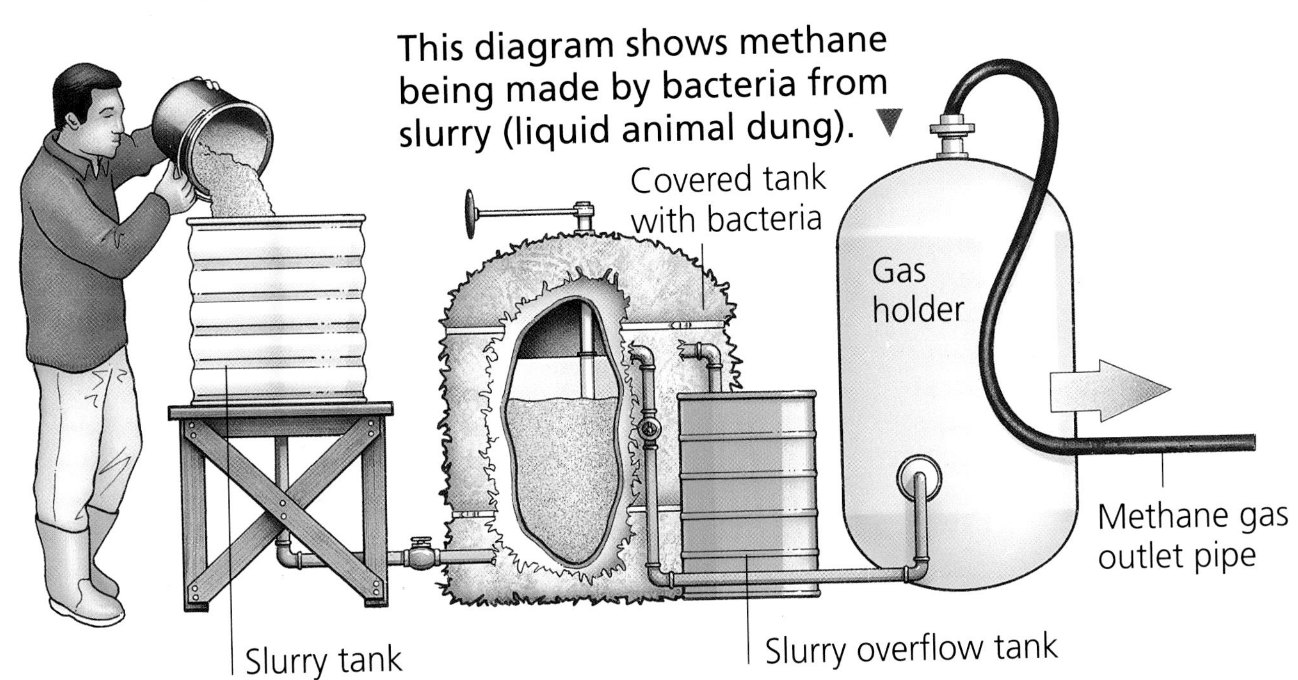

Problems with bio-energy

Peat bogs take thousands of years to form. So digging peat from them destroys the peat bog.

Trees can grow fairly quickly. But in many places people are cutting trees down and burning them more quickly than new trees can grow. And stripping trees from the land leaves the earth bare. Then wind can blow the earth away. Heavy rain can wash the earth away. This is happening in many parts of the world such as Africa, India and South America.

As more and more people live on the Earth they use more and more wood. This map shows the places where there is a shortage of wood.

Collecting wood. Burning wood and peat give off carbon dioxide in the same amounts as living trees do. Fossil fuels give off masses of extra carbon dioxide that has been stored in the ground for years as coal, oil and gas. ▼

Global warming

Many people think that the Earth's atmosphere is heating up. This is called global warming. It could be happening because we are burning so many fossil fuels, trees and peat. All these things give off carbon dioxide when they are burnt. The big increase in carbon dioxide could be causing global warming.

Household rubbish sacks about to be buried. When they have rotted, the gas given off can be used. ▼

A geothermal power station in Kenya. ▼

▲ Natural hot water from underground or waste water from a power station can heat greenhouses.

Good things about geothermal power

Geothermal power stations are kinder to the environment.

- They use heat from underground which is renewable (will not get used up).
- They produce little or no harmful gases.
- They produce little or no harmful waste materials.

FACTFILE

The geothermal power station on the island of Hawaii produces as much power as 79.5 million litres of oil would produce. The island does not need so much oil brought to it in large oil tankers.

Bad things about geothermal power

Geothermal power does have some problems. Geothermal power stations take water from the ground. This could make the natural level of water in the ground become too low. The ground would dry out and shrink. Then land would start sinking. So water has to be pumped back underground all the time.

And underground gases can escape. They can cause air pollution, and be very noisy.

Methane gas being burnt. Burning methane gas does not give off poisonous gases or pollute the air.

▲ A bubbling, hot mud pool in Iceland.

▲ There are natural hot springs in Bath, in England, from an old volcano. The Romans built bath houses near the hot spring.

Hot springs

Water can be heated by hot underground rocks. When the water comes to the surface it forms hot springs. These are common in volcanic areas such as Iceland and New Zealand.

In some places fine dust mixes with the hot springs. This forms bubbling hot mud pools.

The Romans

The Romans discovered hot springs in Baden-Baden in Germany, and built baths there. They not only bathed in the water, they piped the hot water from the springs to heat all their bath buildings.

Iceland

Today, some buildings in the capital city of Iceland are heated by water piped from nearby hot springs.

FACTFILE

In 1883, workers were building the Canadian Pacific Railroad. They saw a huge column of smoke in the distance. When they got nearer they found it was not smoke but steam. It came from boiling hot underground water. This is called a geyser. Soon people came to see it, and other geysers.

◀ There are many hot springs in the Yellowstone National Park in Wyoming, in the USA. Here water flows down a hill. Brightly coloured algae grow in some of the pools.

FACTFILE

TALLEST GEYSERS
The tallest at present:
Steamboat Geyser, Yellowstone National Park, USA (115 metres)

The tallest ever recorded:
Waimangu Geyser, New Zealand (460 metres)

A geyser on the island of Lanzarote.

USING GEOTHERMAL AND BIO-ENERGY

How a geyser works

Water trickles down a crack in the ground. It falls on to the red-hot rocks underground. The water boils and flashes to steam. The water and steam are forced up, and meet more cold water coming down. Now all the water and steam is blasted out of the ground in a huge jet.

Where geysers are in the world

Most geysers are in New Zealand, Iceland, the USA, and Russia.

The word geyser comes from Geysir in Iceland. Geysir is one of the most spectacular geysers in the world. Every five to thirty-six hours there is a jet of steam and water 60 metres high.

▲ Yellowstone National Park, in the USA, has more geysers close together than in any other part of the world.

Yellowstone National Park

This is the oldest national park in the world. The park land is full of old volcanoes. The rocks below the ground are very hot. This means there are many hot springs, boiling mud pools and geysers. The most famous geyser is Old Faithful, which erupts regularly.

▲ Thousands of tourists visit Yellowstone National Park every year, to watch Old Faithful in action.

▲ Deep underground, water is flowing down cracks towards the red-hot rocks. The heat will flash the water to steam and blast it out of the ground.

Old Faithful

Some facts about this geyser:

- It erupts every 37-93 minutes.
- It can shoot up to 52 metres high.
- It blasts about 40,000 litres of water into the air.

Geothermal power stations

Most of the world's geothermal power stations are in the USA, Mexico, Japan, Russia, New Zealand and Italy. But other countries are beginning to build them too.

How the underground heat is used

There are three main ways that underground heat is used.

- Steam from underground directly turns turbines.
- Underground hot water is changed to steam by lowering air pressure and then the steam turns the turbines.

Pipes to stop a road freezing in Iceland. Waste hot water from a nearby geothermal power station can be used to heat everything from greenhouses to road surfaces. ▼

- Underground warm water is used to heat another liquid that boils at a lower temperatrue and the vapour from this turns the turbines (see the diagram below).

FACTFILE

There are 45 geothermal power stations in the USA. They make electricity for 3.5 million people. If all known geothermal sources in the USA were used, this would make 27 times as much energy as the whole USA needs every year.

A BINARY PLANT
When the underground water is hot (100 -175 degrees Celsius) it is used to heat another liquid with a lower boiling point than water. Vapour from this liquid drives the turbines. ▼

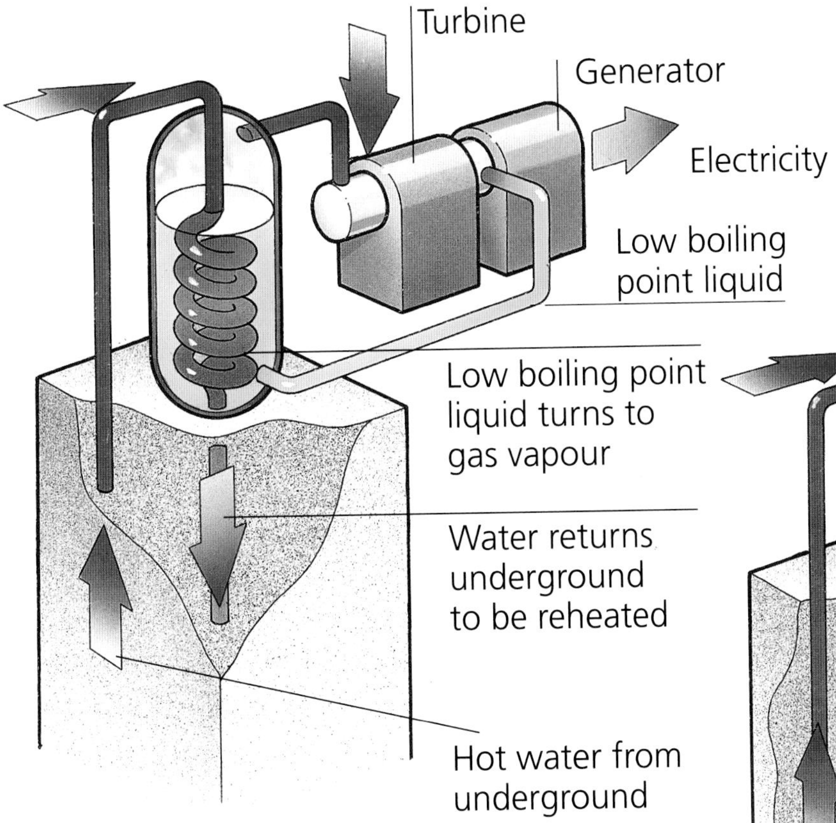

Turbine
Generator
Electricity
Low boiling point liquid
Low boiling point liquid turns to gas vapour
Water returns underground to be reheated
Hot water from underground

A FLASH PLANT
When the underground water is very hot (175 degrees Celsius or more) it is piped to the surface and flashed to steam (by lowering the pressure). Then it turns the turbines. ▶

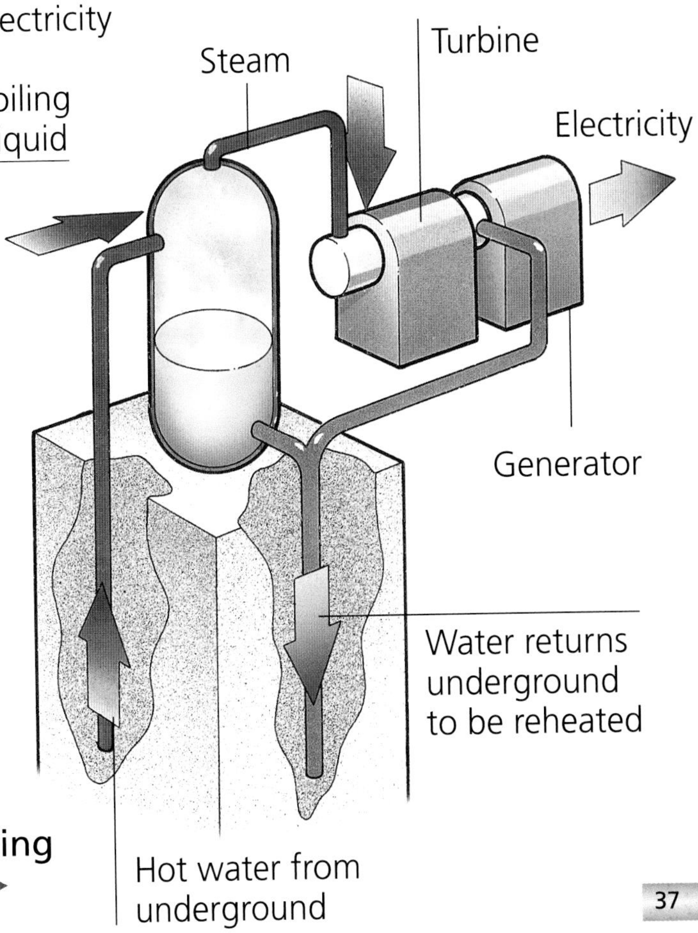

Steam
Turbine
Electricity
Generator
Water returns underground to be reheated
Hot water from underground

The Geysers geothermal power station

The Geysers is built in a volcanic area in California. It first produced electricity in 1960. By 1986, this power station was supplying more than a million people with their electricity. At full power it can supply nearly all the electricity needed by San Francisco.

This geothermal power station called The Geysers supplies electricity to people in northern California, in the USA. ▼

The Geysers' steam

The rocks under The Geysers are very, very hot. This means that the power station can use steam straight from the ground.

When the steam has been used it cools down and becomes water. The water is injected back into the ground to heat up again.

▲ Drilling to find hot rocks underground.

At The Geysers site, steam comes up from underground all the time. ▼

▲ A car in Brazil is filled with gasohol. Gasohol is made from ordinary petrol and liquids produced from cassava and sugar cane.

Plants, wood and rotting rubbish

Both plants and rotting rubbish can be used to make liquid or gas fuels. A rubbish site begins to produce gas after about three years. It is better if the plant rubbish can be separated from other rubbish. Plant rubbish rots down faster by itself.

Heating wood without much oxygen gives off gases. These gases can be burnt inside a gas-turbine engine. The engine's hot gases then heat water in a boiler. This heats water and makes steam. The steam drives the turbines. The turbines drive electricity generators.

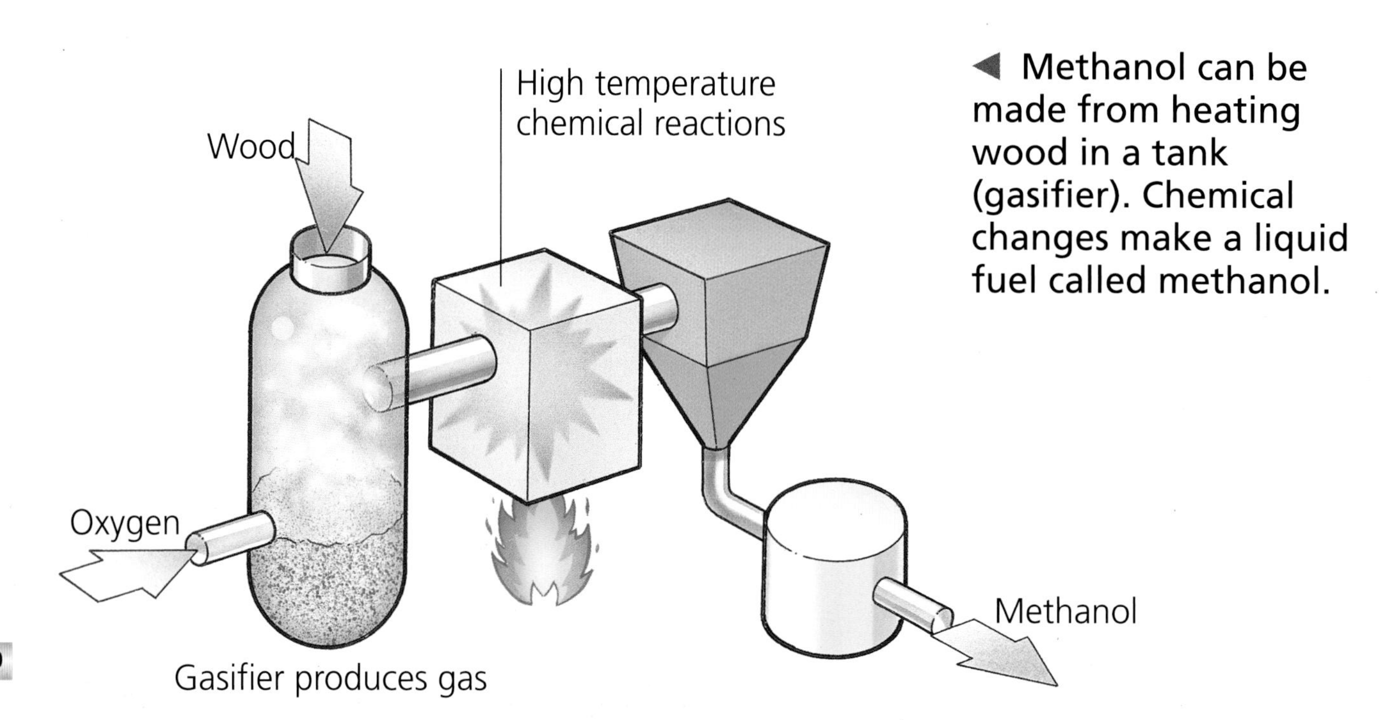

◀ Methanol can be made from heating wood in a tank (gasifier). Chemical changes make a liquid fuel called methanol.

Each sunflower produces a little oil. Thousands of sunflowers can produce enough oil to be a useful fuel.

THE FUTURE

FACTFILE

In the future geysers may be harnessed to make electricity. Geothermal energy and other sources of energy such as the Sun, wind and waves do not get used up (they are renewable). They will become more important than coal, oil and natural gas.

The future of geothermal power

All countries would like to have energy sources that will not run out. By the end of the twenty-first century, the USA could be getting about one-third of its electricity from geothermal power stations.

Geothermal power stations, like this one in New Zealand, may become more common in the future.

Geyser power

It may become possible to use the huge jets from geysers to power turbines directly. It may even be possible to make artificial geysers. This would mean drilling holes in rocks and injecting water into the red-hot rocks below.

People bathe in hot mud pools on an island near Italy. Many people find this helps to ease muscle pain. ▼

The future of bio-energy

Fuels from plants can be sold to other countries. This would be very useful for poor countries. (Geothermal energy can only be used by the country where volcanoes and hot rocks are found.)

Different sorts of plants for fuel

Many different sorts of plants could be used for fuel.

- Kenya is growing sugar to make into fuel.
- Miscanthus is a fast-growing grass. Countries in the Far East are looking for ways of making it into a fuel.
- In California, in the USA, seaweed has been used to make a gas fuel.

Future filling stations

In the future, vehicles may run on different types of fuel. Filling stations may sell gasohol, methanol and biogas. These are all fuels made from plants and animal waste.

This picture shows a site using renewable energy sources such as wind power, solar panels and biogas digesters. In the next hundred years, more of our energy needs will be met in this way. ▶

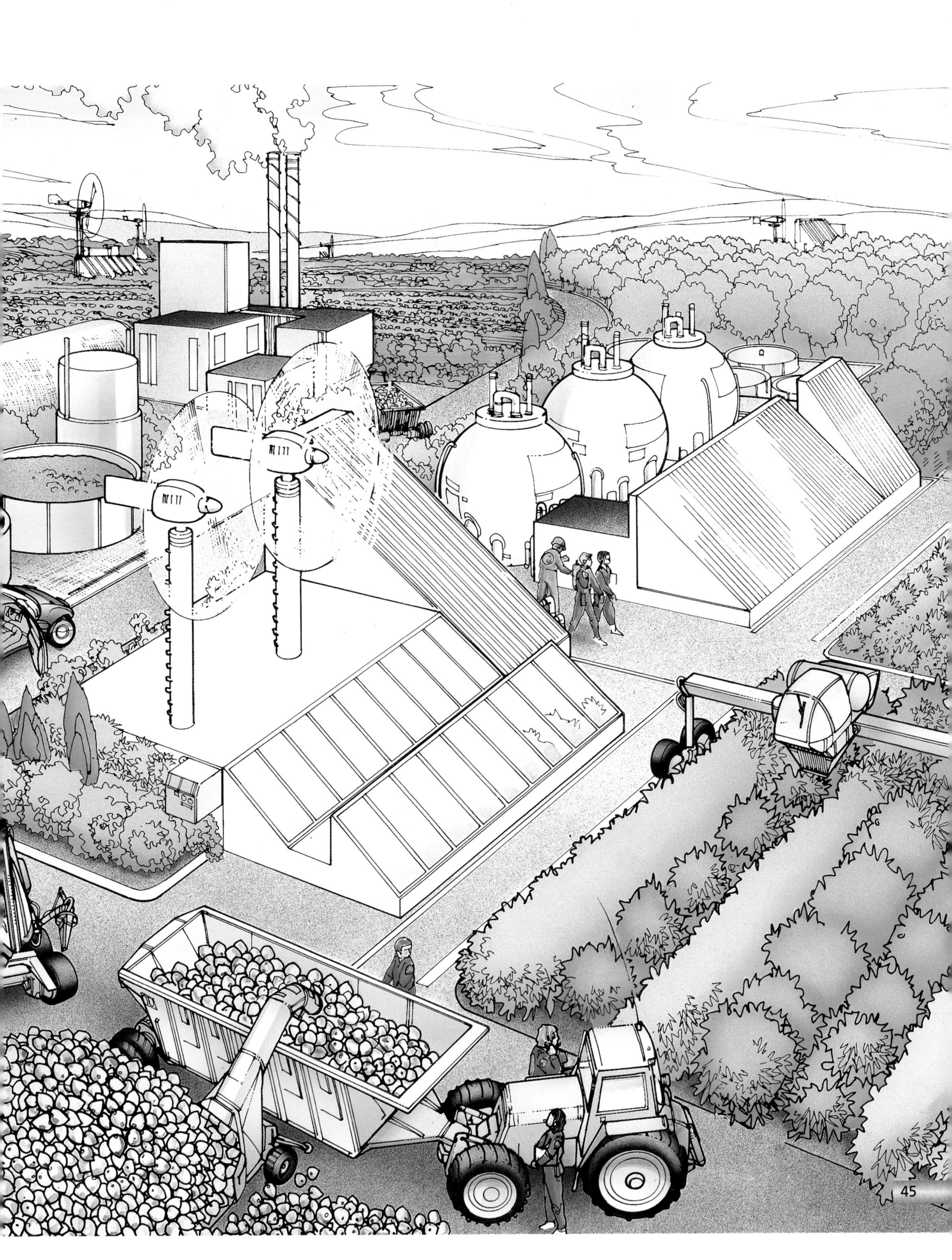

GLOSSARY

Air pollution Dirty air.

Alcohol A liquid that burns easily.

Bacteria Tiny creatures that cannot be seen by the naked eye.

Bio-energy Power produced from plants and animals.

Bio-gas Gas from animal or plant sources.

Carbon dioxide A gas that is present in the air.

Energy The ability to do work.

Fossil fuel Coal, oil or natural gas formed over millions of years from the remains of plants and animals.

Fuel A material that is burned to release the energy that is stored in it.

Gasohol A mixture of alcohol and petrol.

Generator A machine to produce electricity.

Geothermal Heat from the Earth.

Global warming The warming of the Earth's atmosphere.

Joule A unit of energy..

Kilowatt One thousand watts.

Megawatt One million watts.

Methane A gas that burns easily. It can be made from rotting material.

Methanol A type of alcohol.

Molten Melted, as in molten rock - so hot that it is liquid.

Natural gas Gas usually found deep underground together with oil deposits.

Power station A building where energy from a fuel is used to make electricity.

Turbine Blades like a giant fan turned by gas or liquid.

Vapour When a liquid is heated to boiling point it turns to gas. This is called vapour.

Watt A measurement of energy. One watt equals 1 joule being used in 1 second.

Watt-hour One watt being used for one hour. A 40-watt light bulb being used for one hour uses 40 watt-hours of energy.

FURTHER INFORMATION

Books to read

Action for the Environment: Energy Supplies by Chris Oxlade and Rufus Bellamy (Franklin Watts, 2004)

Alpha Science: Energy by Sally Morgan (Evans, 1997)

A Closer Look at the Greenhouse Effect by Alex Edmonds (Franklin Watts, 1999)

Cycles in Science: Energy by Peter D. Riley (Heinemann, 1997)

Essential Energy: Energy Alternatives by Robert Snedden (Heinemann, 2002)

Future Tech: Energy by Sally Morgan (Belitha, 1999)

Saving Our World: New Energy Sources by N. Hawkes (Franklin Watts, 2003)

Science Topics: Energy by Chris Oxlade (Heinemann, 1998)

Step-by-Step Science: Energy and Movement by Chris Oxlade (Franklin Watts, 2002)

Sustainable World: Energy by Rob Bowden (Hodder Wayland, 2003)

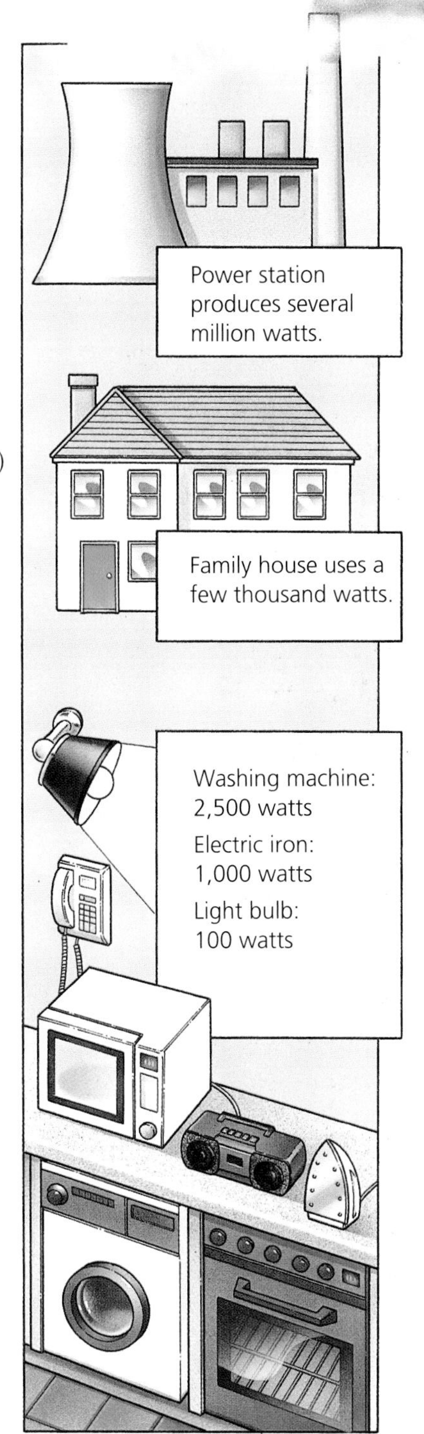

ENERGY CONSUMPTION

The use of energy is measured in joules per second, or watts. Different machines use up different amounts of energy. The diagram on the right gives a few examples. ▶

INDEX